Instructor's

to accom

Standard and Microscale Experiments in General Chemistry

Third Edition

Carl B. Bishop
Clemson University

Muriel B. Bishop
Clemson University

Kenneth W. Whitten
University of Georgia

Saunders College Publishing
Harcourt Brace College Publishers

Fort Worth Philadelphia San Diego New York Orlando Austin
San Antonio Toronto Montreal London Sydney Tokyo

Printed in the United States of America.

Bishop, Bishop & Whitten; Instructor's Manual to accompany the Lab Manual for General Chemistry and General Chemistry with Qualitative Analysis, 5E.

ISBN 0-03-015633-5

567 280 987654321

PREFACE

Chemistry is an experimental science. What we know about chemical systems comes from experimental observations made by thousands of scientists. Theories provides a framework for understanding and explaining experimental observations and a basis for designing new experiments.

We have attempted to provide an opportunity for students to learn to make measurements and observations, and to interpret the data from these measurements and observations. Some of the earlier experiments in this manual provide an introduction to basic laboratory techniques used in all chemistry laboratories. These techniques will also be used in later experiments. The equipment which chemists use may be more sophisticated, but the principles in simple equipment used in introductory laboratories is the same.

Written descriptions are illustrated by art work that is large and clearly drawn so that students can "see for themselves."

The appropriate chapters, sections, and topics in the text reference, and in the manual are listed at the beginning of each experiment. Students can find the corresponding sections in other general chemistry books with little effort by referring to the topic names.

The rather detailed review of the relevant theory and principles given in the *Introduction* to each experiment will help students associate theory with the experiment, and help laboratory instructors prepare for preliminary lectures. Procedures are described in stepwise detail so that the student can set up the necessary apparatus correctly and quickly.

Safety precautions are emphasized at appropriate points throughout the manual. We find it useful to require the student to perform the safety exercise (pp. 1-7) during the check-in laboratory, and to take the safety quiz (p. 5). Frequent reference to these safety measures directs attention to the appropriate hazards in a given laboratory and the safety measures required.

Each experiment has a set of *Prelab Questions* designed to check for understanding of the principles and procedures involved in the experiment. The laboratory instructor may choose to use these prelab questions as 'admission tickets' to the laboratory. A set of *Post Lab Questions* follows the Report Form. These questions check for understanding of factors that affect the experimental results. These are designed to be answered during the experiment.

We have provided Report Forms with space for observations and interpretations. Since most experiments also have an "unknown," we have provided a space for the unknown number close to the result to facilitate grading.

The *laboratory directions* and a brief description for the experiments in qualitative analysis are included in this manual so that students do not have to bring large (expensive) textbooks into the laboratory. The steps are those found in the qualitative analysis chapters in *GENERAL CHEMISTRY WITH QUALITATIVE ANALYSIS*, 5th edition, by Whitten, K. W., Davis, R. E., and Peck, L., published by Saunders College Publishing (1996). A minimum amount of discussion is included in each experiment. However, the student must prepare

for the laboratory by reading the chapters in the text in order to fully grasp the chemist
involved in each step.

We have included a reasonable number of quantitative and descriptive experiments so that each laboratory director has considerable flexibility in the choice of experi ments . The experiments may be divided into classes as shown below.

Experiments	Classifications
1, 2	Basic laboratory techniques (simple measurements and observations)
3, 4, 5	Gravimetric techniques
6	Introduction to inorganic synthesis
7, 8, 9, 10	Introduction to reaction chemistry (organization and interpretation of data)
11, 33	Structure and bonding
12, 13, 14,15	Introduction to physical measurements
16, 17, 18	Acid-base chemistry (titrations)
19	Oxidation-reduction reactions
20, 21, 22, 23, 24, 25, 26	More sophisticated physical measurements
27, 28, 29	"Descriptive" experiments and more inorganic synthesis
30, 31, 32	Introductory organic experiments
34, 35, 36, 37, 38	Introduction to Qualitative Analysis

Some **microscale experiments** that require small amounts of chemicals and small equipment are listed below:

7, 8, 9, 10, 19, 21, 23, 28, 30, 33, and the qualitative experiments (34-38).

TO THE LABORATORY INSTRUCTOR

A training program on safety in the chemistry laboratory is an absolutely essential part of preparing laboratory instructors for supervision in undergraduate chemistry laboratories. Laboratory safety training materials are available from the American Chemical Society. We suggest that materials such as "Starting with Safety", the 35 minute videtape developed by Lois Wickstrom, be used the first laboratory period. A formal section on safety must be presented at each orientation meeting of the lab director with laboratory instructors, and at the pre-lab meeting of laboratory instructors with students, in order to promote an awareness of lab safety.

This manual has used updated references, but the laboratory director is responsible for updating safety standards for the conditions in their own laboratories. It is the responsibility of the laboratory instructor to point out hazards to students. Laboratory instructors and students should have access in the laboratory to reference documents on safety such as the following:

(1) Safety in Academic Chemistry Laboratories, Department of Educational Activities, ACS, 1155 16th St. N.W., Washington, D.C. 20036.

(2) NIOSH Publications No. 78-104A, 104B and 77-205 on Toxic Effects of Substances and Carcinogens. Superintendent of Documents U. S. Government Printing Office, Washington D. C. 20402.

(3) Material Safety Data Sheets (MSDS)

(4) Safe Handling and Disposal of Chemicals, J. T. Baker Inc., 222 Red School Lane, Phillipsburg, NJ 08865

The *Instructor's Manual* provides detailed lists of chemicals, equipment, the amount of time required for each experiment and suggested answers to the pre- and post- lab questions. Copies are available to all adopters of this laboratory manual.

About 40 percent of the experiments in this manual are micro-scale. Experiments in which learning the use of standard, "real", quantitative equipment or techniques is a major objective, remain standard. However, chemicals that create a serious disposal problem have been decreased in quantity. If the chemicals are non-hazardous, can be dumped down the sink or are relatively cheap, then standard amounts have been suggested.

CONTENTS

EXPERIMENT 1
DENSITY OF SOLIDS AND LIQUIDS:
MEASUREMENTS AND SIGNIFICANT FIGURES

1. Important Reminders

a. Even though the concepts in this experiment are simple students learn first hand that chemistry is a quantitative science. They learn how to use the balance and volumetric equipment. The experiment shows them the application of the terms, accuracy, precision, significant figures, and also the importance of correctly reporting data The limitations of actual measurements versus the number of digits produced on a calculator must be stressed in this first experiment.

2. Special Apparatus and Equipment

	Per Student
Metric ruler	1
Thermometer (to read to 0.1 °C)	1
Pipet, 10 mL and bulb	1
10 mL graduated cylinder	1
25 mL graduated cylinder	1
Plastic micropipet (Beral)	2
Razor blade to cut micropipet	2 per class
Crucible tongs	1
Objects for density determination (small enough to fit in graduated cylinder)	
Glass marbles or steel ball bearings	1
Glass rods (20 mm length x 10-12 mm diam.)	1
Metal rods (Al, Cu, steel, brass, etc.) square or round	1
Broken pieces of coal, stone, lump sulfur, etc. (volume by displacement of water)	1

3. Chemicals

solutions	Per 100 Students
NaCl (for unknown)	about 200 g (3000 mL)

4. Unknowns

a. Issue approximately 30 mL of an unknown to each student.

b. The following unknowns (A, B, C, D) are prepared by carefully weighing out the "g NaCl", transferring the NaCl to a volumetric flask, and diluting to the mark.

Typical Salt Solutions for Unknowns

Unknown	% NaCl by mass	Density	g NaCl	g H_2O
A	2.00%	1.0143 g/mL	20.25	994.01
B	5.00%	1.0359 g/mL	50.80	984.11
C	7.50%	1.0541 g/mL	79.06	975.04
D	10.00%	1.0726 g/mL	107.26	965.34

c. If you use a 1.000 liter volumetric flask, these masses of NaCl will prepare one liter of solutions of the listed densities.

d. The grading scale can be based on these densities to the third decimal.

EXPERIMENT 2
SEPARATION OF THE COMPONENTS OF A MIXTURE

1. Important Reminders
   a. Observe and call students' attention to all cautions. Remind them that they must wear departmentally approved eye protection at all times while they are in the laboratory.

   b. The value of this experiment comes in the development of quantitative techniques. A discussion of how poor techniques (spattering spills, incomplete washing, etc.) contribute to error will help the student recognize and practice good laboratory techniques. Discussions of the practical application of separation procedures in everyday chemical processes such as water purification and sewage treatment will make this experiment more interesting.

2. Special Apparatus and Equipment

	per student
150 mL beaker	1
Wire gauze	1
Iron ring	1
Tongs	1
Filter paper	5-6 pieces
Evaporating dish	1
Watch glass	1
Filter funnel	1
A small scoop or spatula (for use by instructor when weighing unknowns)	1

3. Chemicals

	Per 100 Students
SiO_2	125 g
NaCl	125 g
NH_4Cl	125 g

4. Unknowns
   a. Each student receives an unknown. The unknown number is recorded by the student's name.

   b. Each unknown must be prepared and weighed individually. A small scoop or spatula can be used to add 0.8 - 1.2 g of each of the three compounds into a weighing flask. The total weight after each addition is recorded. The total sample (2.5 - 3.5 g) can be poured from the spout into a clean test tube, stoppered, and numbered.

   A weighing flask can be constructed by modifying a bottle or flask to give a flat bottom, pouring neck (A) and an opening (B) through which the sample can be added during weighing (see diagram).

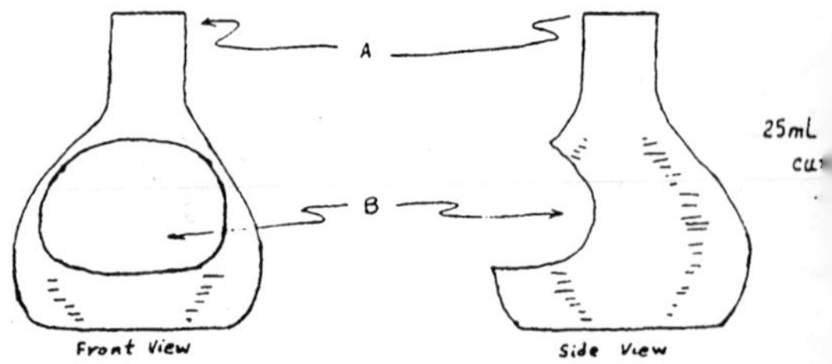

EXPERIMENT 3
INDIRECT GRAVIMETRIC DETERMINATION OF A HYDRATED COMPOUND IN A MIXTURE

1. Important Reminders

a. Observe and call students' attention to all cautions. Remind them that they must wear departmentally approved eye protection at all times that they are in the laboratory.

b. It is very difficult to mix two solids together and get a uniform mixture; therefore, each unknown should be made by mixing the quantities of known weight as described in 4.b. in experiment 2.

c. Obtaining a constant weight on the crucible and lid requires heating to red heat and cooling slowly. This step is time consuming and should be set up as soon as possible upon going to lab in order to finish during one period.

2. Special Apparatus and Equipment

	per student
Small vials or test tubes to issue unknown	1
Clay triangle	1
Crucible and cover	1
Tongs	1

3. Chemicals

	Per 100 Students
$BaCl_2 \cdot 2H_2O$ (unknowns)	145 g $BaCl_2 \cdot 2H_2O$
NaCl (unknowns)	200 g NaCl

4. Unknowns

Typical Unknowns	#g $BaCl_2 \cdot 2H_2O$	#g NaCl
A. 40.0% $BaCl_2 \cdot 2H_2O$ and 60.0% NaCl	1.20	1.80
B. 45.0% $BaCl_2 \cdot 2H_2O$ and 55.0% NaCl	1.35	1.65
C. 50.0% $BaCl_2 \cdot 2H_2O$ and 50.0% NaCl	1.50	1.50
D. 55.5% $BaCl_2 \cdot 2H_2O$ and 50.0% NaCl	1.65	1.35

Weigh and number each sample as in Experiment 2.

EXPERIMENT 4
PERCENT COPPER AND FORMULA WEIGHT OF A COPPER COMPOUND

1. Important Reminders
   a. Observe and call students' attention to all cautions. Remind them that they must wear departmentally approved safety glasses at all times that they are in the laboratory.

   b. Dispense 6 *M* HCl, 95% ethanol and acetone from self-dispensing bottles (such as Brinkman Dispensette, Asko). Label bottles with NFPA hazard codes.

   c. Have baking soda available to neutralize acid spills.

2. Special Apparatus and Equipment

	per student
Drying lamps or ovens	2 per class
Small vial (or test tube) to issue unknowns	1
Filter paper	2-3 pieces
Beaker, 150 mL	2
Rubber policeman for stirring rod	1
Wash bottle	1
Watch glass	1

3. Chemicals

	Per 100 Students
6 *M* HCl	1500 mL
Mg turnings	100 g
95% alcohol (flammable)	1500 mL
acetone (flammable)	1500 mL
filter paper	250 sheets

4. Unknowns
   Students will give you a clean dry test tube in exchange for a numbered unknown tube which contains approximately 4g of one of the following mixtures*.

Copper compounds	Per 100 Students	Per Unknown
A. $CuSO_4 \cdot 5H_2O$	100 g*	4 g
B. CuO	62.5 g*	2.5 g
C. Cu $(CH_3COO)_2 \cdot 2H_2O$	100 g	4 g
D. Basic cupric carbonate (analysis on label)	100 g*	4 g

If analytical balances are available, the amount of unknown issued can be reduced from 4 g to 2 g . The concentration of HCl can be reduced to 3*M* and Mg turnings can be reduced to 0.4 g without seriously affecting accuracy.

* Assumes four compounds are used for unknowns, and 25 students are given each unknown.

EXPERIMENT 5
A SEQUENCE OF CHEMICAL REACTIONS

1. Important Reminders

a. Observe and call students' attention to all cautions. Remind them that they must wear departmentally approved eye protection at all times that they are in the laboratory.

b. Dispense 6 *M* HNO_3, 4 *M* NaOH, 6 *M* HCl, 6 *M* NH_3, 6 *M* CH_3COOH, 6 *M* H_2SO_4, 95% ethanol and acetone from self-dispensing bottles (such as Brinkman Dispensette, Asko). Label bottles with NFPA hazard codes.

c. Have baking soda available to neutralize acid spills.

2. Special Apparatus and Equipment

Filter paper and filter apparatus (ring and stand)	1 per 2 students
500 mL beaker	1
Drying lamp or oven	1 per class

3. Chemicals

	Per 100 Students
copper wire	305 g
6 M HNO_3	1000 mL
4 *M* NaOH	1500 mL
6 *M* CH_3COOH	300 mL
phenolphthalein (1% in 95% ethanol)(dispense from dropper bottles)	50 mL
6 M HCl	1200 mL
6 *M* aqueous NH_3	1500mL
6 *M* H_2SO_4	2000 mL
magnesium turnings	50 g
95% ethanol (flammable)	2500 mL
acetone (flammable)	2500 mL
(NOTE: *If Steps 17-21 are omitted by going from step 16 to Step 22 , reduce the amount of ethanol and acetone by one-half.*)	
litmus paper, red	4-5

4. Unknowns

a. Weigh a 100 cm length of 20 or 22 gauge copper wire. Calculate the lengths that have the following weights and then carefully cut these lengths.

0.220 g   0.240 g   0.260 g   0.280 g   0.300 g

b. These samples can be issued as "unknowns."

c. The laboratory instructor should verify the balance readings and initial the student values in Steps 2, 14, 19, 36 and 40.

d. As a safety precaution, the laboratory instructor may dispense the 6 *M* HNO_3 , HCl, and H_2SO_4 in the hood.

EXPERIMENT 6
THE PREPARATION OF COMMON ALUM FROM SCRAP ALUMINUM

1. Important Reminders

a. Observe and call students' attention to the hazards involved in the use of strong bases and acids and flammable solvents. Remind them that they must wear departmentally approved safety glasses at all times that they are in the laboratory.

b. Dispense 2 *M* KOH, 6*M* H_2SO_4, and 95% ethanol from self-depensing bottles (such as Brinkman Dispensette, Asko). Label bottles with NFPA hazard codes.

c. Have baking soda available to neutralize acid spills.

d. In your discussion you should point out that the method by which aluminum is recovered from aluminum cans involves heating the scrap aluminum to very high temperatures. Chemical processes use less energy but they have not been as well accepted so far. Many towns have aluminum recovery industries. A field trip or video tape of such site might be an added point of interest.

2. Special Apparatus and Equipment

	Per Student
Wash bottle	1
Glass wool (small plug)	1
Buchner funnel	1
Vacuum filtration assembly and aspirator	1
Heat lamps or drying ovens	1 per 10 students
Sand paper (5 cm X 5 cm) if scrap beverage cans are used	1
Filter paper	2-3 pieces

3. Chemicals

	Per 100 Students
Scrap aluminum can or Al foil	80 g
2 *M* KOH	4000 mL
methyl red indicator(dispense from dropper bottles)	50 mL
6 *M* H_2SO_4	1000 mL
95% ethyl alcohol	2500 mL

4. Unknowns

a. If scrap beverage cans are used, cut off the ends of a can and cut down one side to produce a flat sheet. Weigh the sheet and calculate the size of a 1 g sample. Cut the sheet into samples of about 0.9 g.

b. If aluminum foil is used, weigh a large sheet and calculate the size of a 0.9 g sample. Cut pieces of these approximate dimensions for student samples.

EXPERIMENT 7
SOME REACTIONS OF METAL IONS
(A Micro-scale Experiment)

1. Important Reminders

a. Observe and call students' attention to hazards involved in using the substances required in this experiment. Remind them that they must wear departmentally approved eye protectiion at all times that they are in the laboratory.

b. Dispense 4 *M* NaOH, and 6 *M* NH_3 from self-dispensing bottles (such as Brinkman Dispensette, Asko). Dispense petroleum ether from a small bottle in the hood--away from flames. Label bottles with NFPA hazard codes.

c. Have baking soda available to neutralize base spills.

d. Label the chemicals with concentrations and FULL FORMULA (i.e., 0.1*M* $Pb(NO_3)_2$) rather than as Pb^{2+}. This emphasizes the fact that although they are testing for the presence of the positive metal (Pb^{2+}) ion in solution, the oppositely charged non-metal ion (NO_3^{1-}) is also present but does not participate in the chemical test which is specific for the metal ion.

e. The laboratory report sheets are structured in such a way as to give order to a tremendous amount of data that accumulates from these tests on metal ions. The laboratory instructor could further aid the student by reminding them that all six tests must show the presence of the ion (positive test) in order for the ion to be correctly identified as present in the unknown. The laboratory instructor might even demonstrate with a real unknown.

f. Provide containers for students to dispose of their waste petroleum ether. Consult MSDS from company for safe disposal techniques and hazards.

2. Special Apparatus and Equipment

	Per Student
24 well test plate	1
Small labels and/or grease pencils	
Micro-sized stirring rods (either glass rods or toothpicks)	10
Medicine droppers(or Beral micropipets)	10

3. Chemicals

a. Metal ions - Solutions of metal ions are dispensed in <u>100 mL dropping bottles</u>. A set (8 dropping bottles) should be available for every 4-6 students.

Chemicals	Per 100 Students
0.1 *M* $Pb(NO_3)_2$	550 mL*
0.1 *M* $Cu(NO_3)_2$	550 mL*
0.1 *M* $Fe(NO_3)_3$	550 mL*
0.1 *M* $Co(NO_3)_2$	550 mL*
0.1 *M* $Zn(NO_3)_2$	550 mL*

	Per 100 Students
0.1 *M* $Mg(NO)_3)_2$	550 mL*
0.1 *M* $Ba(NO_3)_2$	550 mL*
0.1 *M* KNO_3	550 mL*

* Includes amounts to be issued as unknowns

b. Reagents - Solutions of reagents are dispensed in 100 mL dropping bottles.

2 *M* $(NH_4)_2SO_4$	600 mL
1 *M* $(NH_4)_2S$	600 mL
2 *M* NaI (or KI)	600 mL
2 *M* NH_4SCN	600 mL
4 *M* NaOH	600 mL
6 *M* NH_3	600 mL
petroleum ether (flammable)	500 mL

4. Unknowns

Metal ions - Solutions of metal ions are dispensed in 100 mL dropping bottles. A set (8 dropping bottles) should be available for every 4-6 students.

EXPERIMENT 8
CHEMICAL PERIODICITY - THE HALIDE ION
(A Micro Scale Experiment)

1. Important Reminders

a. Observe and call students' attention to hazards involved in this laboratory experiment. Remind them that they must wear departmentally approved eye protection at all times while in the laboratory.

b. Dispense 6 *M* HNO_3, 6 *M* HCl, 0.1 *M* $AgNO_3$, 6 *M* NH_3, and 3*M* H_2SO_4, from self-dispensing bottles (such as Brinkman Dispensette, Asko). Label bottles with NFPA hazard codes.

c. Have baking soda available to neutralize acid and base spills.

d. Label all chemicals with concentration and full formula (e.g., 0.1*M* NaBr + 0.1*M* NaI) in order to emphasize the presence of both the positive metal ion and the negative halide ion.

e. The laboratory report sheet is structured so as to give order to a tremendous amount of data. The laboratory instructor could further aid the student by going through a sample unknown and discussing the logic used to identify the unknown.

f. Provide containers for students to dispose of their waste petroleum ether. Refer to SMDS sheets for information on disposal and hazards.

2. Special Apparatus and Equipment

24 well test-plate	1
100 mL beaker	1
Small labels and/or grease pencils	
Micro-sized stirring rods (1 mm glass rods or plastic toothpicks)	5-6
10 x 7 micro-sized test tubes	5-6
15 x 125 test tubes	3

3. Chemicals

NOTE: 0.1 *M* SOLUTIONS OF THE HALIDES MAY BE USED RATHER THAN 1 *M* SOLUTIONS.

a. Halide ions - Solutions of the halide ions are dispensed in <u>100 mL plastic dropping bottles</u>. A set (8 bottles) should be available for every 4-6 students.

	Per 100 Students
Sol'n A, 1 *M* NaF	500 mL*
Sol'n B, 1 *M* NaCl	500 mL*
Sol'n C, 1 *M* NaBr	500 mL*
Sol'n D, 1 *M* NaI (may substitute KI)	500 mL*
Sol'n E, 1 *M* each in NaBr, KI (or NaI)	150 mL*
Sol'n F, 1 *M* each in NaCl and NaBr	500 mL*
Sol'n G, 1 *M* each in NaCl and KI (or NaI)	500 mL*
Sol'n H, 1 *M* each in NaCl, NaBr and KI (or NaI)	200 mL
*1 *M* NaF and 1 *M* NaCl	200 mL
*1 *M* NaF and 1 *M* NaBr	200 mL
*1 *M* NaF and 1 *M* KI	200 mL

*These quantities include unknowns

The following solids are present only for students to observe

Solid NaF
Solid NaCl
Solid NaBr
Solid NaI (or KI)

b. Reagents - Solutions of reagents are dispensed in 100 mL dropping bottles

	Per 100 Students
2 *M* $Ca(NO_3)_2$	250 mL
0.1 *M* $AgNO_3$	300 mL
0.1 *M* $Fe(NO_3)_3$	300 mL
cyclohexane or petroleum ether	750 mL
6 *M* HCl	250 mL
6 *M* aq NH_3	250 mL
6 *M* HNO_3	200 mL
5% NaClO (use "Clorox")	200 mL
3 *M* H_2SO_4	300 mL
solid $(NH_4)_2S_2O_8$	200 g

4. Preparation of Unknowns

Issue 5 mL of one unknown solution to each student. Suggested unknowns include:

A. NaF
B. NaCl
C. NaBr
D. KI
E. NaBr + KI
F. NaCl + NaBr
G. NaCl + KI
H. NaF + NaCl
I. NaF + NaBr
J. NaF + KI

EXPERIMENT 9
RELATIVE REACTIVITY OF METALS AND THE ACTIVITY SERIES
(A Semimicro scale Experiment)

1. Important Reminders

a. Observe and call students' attention to all cautions. Remind them that they must wear departmentally approved safety glasses at all times that they are in the laboratory .

The laboratory instructor must perform the demonstration of the reaction of sodium metal with water. ***Be aware that the heat of reaction can ignite the hydrogen gas evolved--therefore a small piece of sodium must be used. and the reaction should be performed behind a safety shield.*** The sodium in placed in water with a drop of phenolpthalein to show both the evolution of a gas (H_2) and the production of the base (NaOH).

b. Dispense 3*M* HCl, 0.2 *M* $AgNO_3$, 3*M* H_2SO_4, 3 *M* HCl, 1 *M* $Mg(NO_3)_2$ (6*M* HNO_3), 1 *M* $Al(NO_3)_3$, 1 *M* $Cu(NO_3)_2$, 1 *M* $Zn(NO_3)_2$, 1 *M* $Sn(NO_3)_2$, 1 *M* $Pb(NO_3)_2$,1 *M* $Ca(NO_3)_2$ and 1 *M* $FeSO_4$ from self-dispensing bottles (such as Brinkman Dispensette, Asko). Label bottles with NFPA hazard codes.

c. Both calcium and sodium are reactive. Any left over calcium should be taken up **before** the students test metals in hot water.

d. Have baking soda available to neutralize acid and base spills.

e. Label all chemicals with concentration and full formula (e.g.,1 *M* $Mg(NO_3)_2$ in 6*M* HNO_3) in order to emphasize the presence of both the positive metal ion and the negative halide ion.

f. If both the Experiments 7 and 9 are performed, point out that the tests in Experiment 7 were used to qualitatively identify the presence of the metal ion whereas in Experiment 9 primarily the ease by which the metal undergoes oxidation (loss of electrons) or the metal ion undergoes reduction (gain of electrons) is being established.

g. The Report Form is designed to give order to a tremendous amount of information so that the student can easily set up an activity series.

2. Special Apparatus and Equipment

	Per student
Small labels and/or grease pencils	
Sand paper to clean metals	
Disposal container for waste metals	
Filter paper soaked with kerosene for handling sodium	
Wire gauze, ring and support stand	1
Beaker, 250 mL	1

3. Chemicals

Per 100 Students

a. Metals

sodium (caution) (3-4 pieces small pea sized pieces for demonstration)	
magnesium ribbon (ten 1 cm pieces/student)	50 g
iron nails (2 p) (one nail/student)	150 g
copper strips (1 cm x 1 cm)	*
zinc strips (1 cm x 1 cm)	*
tin strips (1 cm x 1 cm)	*
lead strips (1 cm x 1 cm)	*
The following metals are furnished as foils or sheets	(1-1 inch x1 inch square per student)
copper*	100 g
tin*	100 g
lead*	200 g
calcium turnings (Collect unreacted Ca in a special container after Step 3 before students begin heating other metals since calcium may react violent when heated)	100 g

* If the unreacted pieces are returned to the students, the metal pieces can be re-used by other students and reduce costs.

b. Reagents and Metal Ions (label with full formula)

3 *M* HCl	2 L
1*M* $AgNO_3$	25 mL*
1 *M* $Sn(NO_3)_2$	25 mL
1 *M* $Pb(NO_3)_2$	25 mL*

*Add (caution) about 10 mL of 6*M* HNO_3 to half of the water before adding the salt. This will prevent the precipitation of metal hydroxides or oxides which can form by hydrolysis.

4. Unknowns

None

No Calcium turnings.
No $Sn(NO_3)_2$
No $Ag(NO_3)_2$

EXPERIMENT 10
PAPER CHROMATOGRAPHY
(Semi Micro-scale Experiment)

1. Important Reminders

a. Note and call attention of the students to all cautions including wearing safety glasses even though small amounts of the reagents are used.

b. The student should be warned that the developing solvent level must be below the origin line on the chromatograph, else the dye may be eluted into the solvent.

c. The spots on the chromatograph must be very small. A demonstration of the use of a thin rod or a capillary tube to make spots will aid the student and cut down on waste. Note: Prepared thin layer silica gel on plastic or aluminum sheets (Kodak) may be obtained from Fischer Scientific Co. or E. M. Science. Thin layer chromatography (TLC) requires smaller chromatographs and less solvent.

d. Dispense the developing solvents and the wash solvents from self dispensing bottles (Brinkman Dispensette, Asko SMI, Scienceware).

e. Provide waste bottles for excess solvents. Check the MSDS for proper disposal methods.

2. Special Apparatus and Equipment

	Per Student
Thin Layer Chromatography--	
400 mL beaker	1
Plastic wrap	15 cm x15 cm square
Rubber bands (large enough to go around the top of the beaker)	1
Ruler	1
TLC sheet (7 cm x 8 cm)	1 sheet
pencil	1
Small glass rod	1
Column Chromatography	
9 inch disposable glass (Pasteur) micropipet	1
Small plug of glass wool	1
Fine sand	
Silica gel	850 mg
One hole #6 or #8 rubber stopper to support pipet	1
3 cm of 1mm capilary tubing*	1
Pinch or screw clamp	1

*Cut the tip of a Beral pipet into 3 sections and use this for capilary tubing. Break off about 5 cm of the glass tip on the glass micropipet so that the capilary tubing will fit tight and no long piece of glass will be inside the plastic.

3. Chemicals — Per 100 Students

Thin Layer Chromatography--

a) Dye solutions- Label with full name. Dispense solutions in 50 mL dropping bottles so as to prevent contamination and waste. Students will take a small amount in the bottom of a very small test tube

Fifty (50) mL of the dye standard solutions, which also serve as *unknowns*, are made up as 3 mg/mL in ethanol, except for methyl red (1 mg/mL).

Dye	Color	R_f
Fluorescein	yellow	0.77
Bromocresol green	yellow-green	0.67
Methyl red	red	0.58
Basic fuchsin	purple-red	0.41
New methylene blue	blue	0.12

(These values are given in Reynolds, R.C. and O'Dell, C.A., *J. Chem. Ed.* **1992,** 69, 989-991)

Developing solvents--

2-propanol/glacial acetic acid (15:2 v/v)

1-propanol/water/acetone (1:1:1: v/v/v)

1-propanol/water/glacial acetic acid (1:1:1: v/v/v)

Chemicals

Fifty (50) mL of the dye standard solutions, which serve as unknowns, are made up as 1 mg/mL in ethanol and dispensed from dropping bottles, Students should take only a few drops in the bottom of a very small test tube.

Dye	Color	Time of elution
Fluorescent	yellow	14-16 min.
Basic fuchsia	purple-red	---
New ethylene blue	blue	25-32 min.

(These values are given in Reynolds, ROC. and OED's, C.A., *J. Chem. Ed.* **1992,** 69, 989-991)

Wash column with the solvent--water/1-propanol/ acetone (1:1:1 v/v/v)

1st Eluting solvent--water/1-propane/ acetone (1:1:1 v/v/v)
and 2nd Eluting solvent--water/1-propanol/ glacial acetic acid (1:1:1 v/v/v)

EXPERIMENT 11
MOLECULAR MODELS, LEWIS DOT FORMULAS
VSEPR THEORY AND VALENCE BOND THEORY
(NO CHEMICALS REQUIRED)

1. Important Reminders

This is a good experiment to use in lecture or as an experiment to begin the second semester (or quarter) during check-in period.

2. Special Apparatus and Equipment

	Per Student
Molecular Model Kit	1 Kit
Plastic foam balls (2 sizes)	8 large, 25 small
toothpicks or pipe cleaners	60 toothpicks or 20 pipe cleaners
wire cutters (if pipe cleaners are used)	1 per 10 students
small nail to make holes (if pipe cleaners are used)	5
protractors	1 per 5 students

Alternatively: any commercial molecular model sets that will allow construction of molecules or ions at the centers of a (an)

(1) straight line
(2) trigonal plane
(3) tetrahedron
(4) trigonal bipyramid
(5) octahedron

3. Chemicals

None

4. Unknowns

None

EXPERIMENT 12
THE MOLECULAR WEIGHT OF A VOLATILE COMPOUND

1. Important Reminders

a. The solvents in this experiment are flammable. Caution the students about the use of these solvents around flames!

b. Remind students that they are required to wear departmentally approved eye protection.

2. Special Apparatus and Equipment

	Per 100 Student
Aluminum foil squares 3 cm x 3 cm	110
600, 800 or 1000 mL beaker	100
250 or 500 mL graduated cylinder (Step 14)	100
125 or 250 mL Erlenmeyer	100
Thermometer (110 °C graduated in 0.1)	100
Straight pins to punch holes in aluminum foil	10
Wire gauze, ring and support stand	100
Utility clamp to hold flask	100
Crucible tongs	100
Barometer	1 per class
Paper towels	

3. Chemicals

	Per Student
Each Unknown volatile liquid (if 125 mL flask used need only 6 mL sample)	12 mL

4. Unknowns

Issue 12 mL of ONE liquid unknown per student.

Suggested volatile liquids for unknown. Caution student about the hazards involved in the use of these substances.

Methanol	Methyl ethyl ketone
Isopropyl alcohol	Ethyl acetate
Cyclohexane	Methyl acetate
Acetone	n-hexane

EXPERIMENT 13
INDIRECT DETERMINATION OF THE MASSES OF PIECES OF MAGNESIUM

1. Important Reminders

a. Note and call to the student's attention all cautions.

b. Dispense 6 *M* HCl from a self-dispensing container such as a Brinkman Dispensette.

2. Special Apparatus and Equipment

	Per Student
10 mL graduated cylinder	1
50 mL burette	1
Burette clamp	1
Meter stick	1 for 10 students
Thermometer (-10 ° to 110 °C graduated in 0.1 °C)	1
Barometer	1 for class

3. Chemicals

	Per 100 Students
6 *M* HCl	1000 L
Mg ribbon	(unknowns below)

4. Unknowns

Measure a 100 cm length of new, bright, magnesium ribbon and determine its mass. Calculate lengths of magnesium ribbon that will produce the following volumes of hydrogen at STP. Carefully cut lengths of magnesium ribbon to produce these volumes of hydrogen.(DO NOT let students use balances.)

Unknown A	25.0 mL H_2 at STP = 0.0271 g Mg
Unknown B	28.0 mL H_2 at STP = 0.0304g Mg
Unknown C	31.0 mL H_2 at STP = 0.0336 g Mg
Unknown D	35.0 mL H_2 at STP = 0.0380 g Mg
Unknown E	38.0 mL H_2 at STP = 0.0412 g Mg

EXPERIMENT 14
STANDARD MOLAR VOLUME OF A GAS

1. Important Reminders

a. Note and call student's attention to all cautions.

b. Dispense 3% H_2O_2 from self dispensing bottles (Brinkman, etc.). Label with proper NFPA hazard codes. 3% H_2O_2 causes stinging, but higher concentrations can cause serious burns and explosion hazards.

b. Reinforce awareness of eye safety if your students are getting lax about wearing safety glasses.

c. If the gas delivery and water delivery systems are preassembled, glass cuts can be minimized and time will be saved.

2. Special Apparatus and Equipment

	Per Student
Equipment for gas delivery and water delivery system (See Figure 14-1, page 168)	
Gallon jug or 2.5 L bottle for assembly (b) (or 1000 mL Florence or Erlenmeyer flask)	1
Beaker, 600 mL	1
250 Erlenmeyer	1
1-hole rubber stopper to fit 250 Erlenmeyer in Assembly (e)	1
2-hole rubber stopper to fit container in Assembly (f)	1
Short pieces of rubber (tygon) tubing (~ 20 cm)	2
Rubber or glass tube tip	1
Pinch clamp for tubing at (g)	1
Graduated cylinder	1
10 x 75 mm test tube	1
Barometer	1 per section
Thermometer	1 for 2 students
Safety shield if higher concentrations of H_2O_2 are used	1 per work station

3. Chemicals

	Per 100 Students
3% H_2O_2 solution (6% gives better results, but can cause burns)	3000 mL
1 *M* $FeCl_3$	750 mL

* The gas delivery and water delivery tubes should be pre-assembled and issued to each student as a unit.

4. Unknowns

none

EXPERIMENT 15
THE MOLECULAR WEIGHT OF A COMPOUND BY FREEZING POINT DEPRESSION MEASUREMENTS

1. Important Reminders

a. Note and call student's attention to all cautions.

b. Have special containers available for collecting the "used benzophenone" and "used acetone." Chlorinated hydrocarbon wastes should be collected in a separate waste container. Do NOT put these waste materials in the sink.

c. Refer to the MSDS sheets for labeling, safety, and safe disposal requirements.

d. A deviation of the molecular weight will normally be ±2% for good results.

2. Special Apparatus and Equipment

	Per Student
400 or 600 mL beaker	1
Notched 1-hole rubber stopper (for thermometer)	1
25 cm length of 18 gauge nichrome wire (to make loop)	1
Test tube (2.5 cm x 25 cm)	1
Thermometer (-10°C to 110°)	1
Clamp for thermometer	1

3. Chemicals

Dispense the following chemicals from dispensing bottles labeled with NFPA Hazard codes.

	Per 100 Students
Benzophenone	2100 g
Acetone	4000 mL

4. Unknowns

Issue about 1.5 g-2.0 g of one organic compound with a molecular weight between 100 and 200 amu which is soluble in benzophenone. Place this unknown in small test tubes, stopper and label.

Some suggested "unknowns" are:

Unknown A	p-dichlorobenzene
Unknown B	benzoic acid
Unknown C	salysilic acid
Unknown D	naphthalene

Check MSDS for safety precautions and disposal.

EXPERIMENT 16
OXYGEN AND ITS OXIDES

1. Important Reminders

a. Note and call students' attention to cautions in this section. 6% H_2O_2 must be dispensed from a self dispensing bottle. It causes severe burns. Label with NFPA hazard codes.

b. A hood with a good flow of air will be needed for Steps 22-25 and 27.

c. The laboratory instructor must dispense reagents such as Na_2O_2, in order that large amounts of the reagent not produce an excessively vigorous reaction.

d. Be aware that **perchloric acid reacts vigorously with organic materials**. The reagent to be used by the students is dilute but **care must be taken in the preparation** of the dilute acid as well as the use of the reagent by the students.

2. Special Apparatus and Equipment

	Per Student
Tongs or tweezers	1
Pneumatic trough	1
Bottles or Erlenmeyer flasks for collecting gas, 250 mL	5
Glass squares (3 in x 3 in), to cover bottle mouths	5
1-hole rubber stopper	1
Large test tube 1" diam.	1
Short pieces of rubber (or tygon) tubing (18 in)	2
2 inch piece of glass tube	1
Universal clamp	1
Deflagrating spoon	1
Porcelain crucible	1

3. Chemicals

	Per 100 Students
1 *M* $FeCl_3$	75 mL
6 % H_2O_2	8 L
3 *M* HCl	3.5L
6 *M* HCl	1.5 L
4 *M* NaOH	1.5 L
0.1 *M* $HClO_4$	250 mL
Al_2O_3	20 g
phorphorus, red	20 g
sulfur	20 g
magnesium ribbon	25 g
charcoal	25 g
calcium, small pea size pieces	25 g
Na_2O_2, rice grain size pieces	25 g
boric acid, rice grain size pieces (labeled "$B(OH)_3$ or H_3BO_3")	25 g
ball steel wool (1/4 -1/2 in diameter)	30 g
charcoal, small piece	20 g

4. Unknowns

none

EXPERIMENT 17
ACID BASE TITRATIONS
ANALYSIS OF ACID SOLUTIONS OF UNKNOWN CONCENTRATION

1. Important Reminders

a. Note and call students' attention to all cautions.

b. Dispense 6*M* NaOH and 6 *M* H_2SO_4 from self-dispensing bottles such as Brinkman Dispensette),

c. Remind your students to use only their own preparations of NaOH and H_2SO_4.

2. Special Apparatus and Equipment

	Per Student
250 mL beaker	1
400 mL beaker	1
Wash bottle	1
Tongs	1
Volumetric pipet, 10 mL, and bulb	1
Burette, 50 mL	1
Stopcock grease (if needed)	
Bottle with cap and stopper, 500 mL	1
125 or 250 mL Erlenmeyer flasks	3
Pipet bulb	1

3. Chemicals

	Per 100 Students
6 *M* NaOH, "Carbonate free"	1000 mL
6 *M* H_2SO_4	250 mL
1% phenolphthalein in 95% ethanol (dispense from dropper bottles)	200 mL
KHP (potassium hydrogen phthalate)(method A only)	2.3 g/student
$H_2C_2O_4{\bullet}2H_2O$ (oxalic acid dihydrate)(method B only)	4.7 g/student

4. Unknowns

Issue 60 mL of an "unknown HCl" solution and 60 mL of an "unknown $HC_2H_3O_2$" solution to each student.

Use a burette to measure the volumes of fresh concentrated acids when preparing these solutions.

Typical HCl Unknown			
A	0.250 *M* HCl	20.32 mL	12.3 *M* HCl solution
B	0.280 *M* HCl	22.76 mL	12.3 *M* HCl solution
C	0.310 *M* HCl	25.20 mL	12.3 *M* HCl solution
D	0.350 *M* HCl	28.46 mL	12.3 *M* HCl solution

Typical CH_3COOH Unknowns

A	4.00% CH_3COOH	4.17 mL glacial CH_3COOH/L solution
B	4.30% CH_3COOH	4.48 mL glacial CH_3COOH/L solution
C	4.60% CH_3COOH	4.79 mL glacial CH_3COOH/L solution
D	5.00% CH_3COOH	5.21 mL glacial CH_3COOH/L solution

EXPERIMENT 18
ACTION OF ANTACID TABLETS

1. Important Reminders

a. Remind the student that there are other factors than the neutralization of stomach acid involved in selecting antacid tablets-- such as for elimination of gases, laxative action, pain relief, etc.

b. The student should use the same NaOH solution throughout the experiment.

2. Special Apparatus and Equipment

	Per student or Pair
Volumetric pipet, 25 mL and bulb	1
Burette, 50 mL	1
Erlenmeyer, 250 mL	1
Bottle for 1 L of NaOH	1
Suction bulb	1
Graduated cylinder, 100 mL	1

3. Chemicals

	Per 100 Students
0.15 *M* NaOH**	15 L
1 antacid tablet***	
0.15 *M* HCl (simulated stomach acid)	17 L
Methyl red indicator (dispense from dropper bottle)	250 mL

* This experiment may be done by pairs of students working together. The quantity of equipment and chemicals would be reduced by half.

** In order to discourage "dry-labbing" and also to insure a sufficient quantity of solution, issue 1 L of 0.15 *M* NaOH solution for each work station. This is approximately 6 grams of NaOH added to each 1 L container and filled with water. At the end of each laboratory period, combine the remaining NaOH solutions to obtain as many full bottles as possible. Prepare new solutions in the empty containers.

*** Use several brands of antacid tablets (Pepto Bismol, Maalox, Rolaids, etc.) so that the students can compare the effectiveness of several products.

4. Unknowns

None

EXPERIMENT 19
OXIDATION-REDUCTION TITRATIONS:
POTASSIUM PERMANGANATE AS AN OXIDIZING AGENT
(Both Standard and Micro-scale Included)

1. Important Reminders

a. Note and call students' attention to all cautions.
b. The standard experiment uses standard equipment--burettes, pipettes, etc. This version is of value to science, engineering, pre-professional health and students in other majors who need "real" laboratory experiences for higher level courses and in their respective careers.
c. The microscale experiment presents the same concepts but uses micro-scale equipment--well plates, Beral pipettes, etc.--not used in industrial or medical laboratories. Great care must be used to get good reproducibility. Small amounts of chemicals are used for the purpose of cutting costs of buying and disposal of chemicals and for safety.
d. Dispense concentrated $KMnO_4$, 2 *M* H_2SO_4 and standard $(COOH)_2$ from self-dispensing bottles such as the Brinkman dispensette.
e. Provide waste bottles for disposal of the $KMnO_4$. Consult the MSDS from the company for proper disposal methods and hazards.

2. Special Apparatus and Equipment

Standard Experiment	Per 1-2 Students
Bottle, 500 mL	1
Burette, 50 mL, and clamp	1
Volumetric pipet, 10 mL, and suction bulb	1
Thermometer (-10°C to 110°)	1
Graduated cylinder, 10-25 mL	1
Micro-Scale Experiment	
24 well plate	2
Plastic micro pipettes (Beral)	4
Micro stirring rods (1 mm glass rods or plastic tooth picks)	4
Razor blade	1

3. Chemicals

Standard Experiment	Per 100 Students
3.00 *M* H_2SO_4	8L
0.050 *M* $(COOH)_2$(oxalic acid)(standardized)	4 L
0.4 *M* $KMnO_4$ (prepare 0.020 *M* $KMnO_4$ fresh from saturated)	2200 mL

Micro-Scale Experiment	Per 100 Students
3.00 *M* H_2SO_4	500 mL
0.050 *M* $(COOH)_2$(oxalic acid)(standardized)	750 mL
0.4 *M* $KMnO_4$ ((prepare 0.020 *M* $KMnO_4$ fresh from saturated)	750 mL
0.1 *M* $MnSO_4$	500 mL

4. Unknowns

$(COOH)_2$

EXPERIMENT 20
HEATS OF NEUTRALIZATION AND HESS'S LAW

1. Important Reminders

a. Safety - although the acids and bases are not especially hazardous under the conditions of the experiment, seye protection is still necessary and good laboratory practice.

b. Good experimental results depend upon Steps 19-20 (repeat of Steps 11-13) being performed quickly. Remind students that they should read ahead and be prepared to carry these steps out quickly.

c. Steps 15-22 are performed for 3 acid/base pairs:
(1) HCl/NaOH (2) CH_3COOH/NaOH (3) $CH_3COOH/NH_3(aq)$

d. Remind students that thermometers are expensive and can be broken easily if they are not supported in the calorimeter or if they roll off the table.

e. Remind students that the 0.1° C calibration allows an estimation in the hundredth's decimal place. Stress that their readings must be recorded to 2 decimal places.

2. Special Apparatus and Equipment

	Per 1-2 Students
Beaker, 250 mL	2
Graduated cylinder, 50 or 100 mL	1
Thermometers (110° C) 0.1° intervals	2
6 oz. plastic foam cups for calorimeter	2
Cardboard or plastic foam lid for calorimeter	1
1/4 inch section of rubber tubing	1
Watch with second hand (a wall clock with a second hand may be used for the whole class)	1
Supports for thermometers	
Pencil	

3. Chemicals

	Per 100 Students
3.00 *M* HCl	5.5 L
3.00 *M* NaOH	11.0 L
3.00 *M* CH_3COOH	11.0 L
3.00 *M* aq NH_3	5.5 L

The above solutions should be standardized to ±0.01 and the concentrations placed on the labels.

4. Unknowns

none

EXPERIMENT 21
RATES OF CHEMICAL REACTION
(A Micro Experiment)

1. Important Reminders

Although small quantities of relatively non-hazardous reagents are used in this experiment, students should be reminded that wearing eye protection in the laboratory is a strict rule.

2. Special Apparatus and Equipment

	Per Pair of students
12 well test plate	1
Plastic micropipet (Beral)	4
Razor blade to cut micropipet at a 45^0 slant	
Watch with second readings (a wall clock with second hand may be used by the whole class)	1
Thermometer	1
Micro stirring rods or toothpicks	4-5
Beaker, 250 mL	2
Beaker, 50 mL	2
Grease pencil	
Test tube rack	1

3. Chemicals

	Per 50 Pairs of Students
0.0400 *M* KIO_3	1 L
0.0200 *M* H_2SO_3* ($NaHSO_3/H_2SO_4$ /starch)	400 mL

*To prepare 1 L of this solution:

(a) grind 17 g of soluble starch in 50 mL of water as a paste.
(b) Slowly add the paste to 900 mL of water while stirring.
(c) Heat the mixture to a gentle boil and stir until clear.
(d) Dissolve 2.40 g of $NaHSO_3$ sodium hydrogen sulfite in the <u>cool</u> solution.
(e) NO EARLIER THAN 12 HOURS BEFORE USE, add 12 mL of 6 *M* H_2SO_4 and dilute to 1 liter.

4. Unknowns

Typical unknowns (10 mL per student pair)

Sample	Concentration	KIO_3/L of solution
A	0.0350 *M* KIO_3	7.490 g
B	0.0380 *M* KIO_3	8.132 g
C	0.0400 *M* KIO_3	8.560 g
D	0.0420 *M* KIO_3	8.988 g
E	0.0440 *M* KIO_3	9.416 g

EXPERIMENT 22
DETERMINATION OF Ka FROM MEASUREMENTS OF pH
(Two Different Methods Offered)

1. Important Reminders

a. Note and call students' attention to all cautions.
b. It is recommended that students work in pairs. If there are not sufficient pH meters, a portion of the class could do this experiment while other groups do other experiments.

2. Special Apparatus and Equipment

	Per Student Pair
Visual Indicator Method	
10 mL volumetric pipet and suction bulb (if 5 mL pipet used then half the volumes)	1
100 mL volumetric flask (or 100 mL graduated cylinder)	5
10 mL graduated (Mohr) pipet (or 10 mL graduated cylinder)	1
"identical" 10 cm test tubes	9
50 mL burette and clamp	1
250 mL Erlenmeyer flask	3
Gum labels	
pH Meter Method	
Beaker, 100 mL	1
Beaker, 250 mL	1
pH Meter (handheld pH meters such as the Corning ChekMite pH Sensor may be used)	1
Glass and standard electrodes (Chek Mite has electrodes)	1
Magnetic stirrer and stirring bar (or stirring rod)	1
250 mL Erlenmeyer flask	3
50 mL burette and clamp	1
5.0 mL pipet and suction bulb	1

3. Chemicals

	Per 100 Students
Visual Indicator Method	
Indicators (dispense from dropper bottles)	
crystal violet	200 mL
thymol blue	200 mL
erythrosine B, disodium salt	200 mL
bromthymol blue	200 mL
methyl orange	200 mL
methyl red	200 mL
0.10 *M* HCl (standardized ± 0.01 *M*)	4 L
0.10 *M* NaOH (standardized ± 0.01 *M*)	8L
Phenolphthalein (dispense from dropper bottles)	400 mL
CH_3COOH solution (approximately 0.1 *M*)	8 L

pH Meter Method

Standard buffer solutions (to standardize pH meter)	
pH 4	1 L
pH 7	1 L
Phenolphthalein solution	400 mL
0.10 *M* NaOH (standardized ± 0.01 *M*)	8 L
Saturated KCl (for glass electrode)	50 mL
CH_3COOH solution (approximately 0.1 *M*)	4 L

4. Unknowns

Typical CH_3COOH unknowns (~100 mL/student):

A	4.00%	CH_3COOH	4.17 mL glacial CH_3COOH/1 L
B	4.30%	CH_3COOH	4.48 mL glacial CH_3COOH/1 L
C	4.60%	CH_3COOH	4.79 mL glacial CH_3COOH/1 L
D	5.00%	CH_3COOH	5.21 mL glacial CH_3COOH/1 L

Unknowns may be prepared in approximately the concentrations listed above. Use the "class average" as the "correct value" for the concentrations of each unknown. EXCLUDE VALUES THAT ARE OBVIOUSLY ERRONEOUS. This method of standardization has been used by us for a number of years. It saves time and money and works well.

	Per Pair Students
Unknown Monoprotic Acids	100 mL
(If only Part A done -	75 mL)

Other Suggested Monoprotic Acid Unknowns:

0.250 *M* HCO_2H
0.250 *M* potassium hydrogen phthalate (KHP)
(KHP acts a monoprotic acid with NaOH)
0.350 *M* HCO_2H
0.300 *M* HCO_2H
0.300 *M* KHP

Other monoprotic acids which could be used are mandelic acid, lactic acid, propionic acid, and glycolic acid.

EXPERIMENT 23
CHEMICAL EQUILIBRIUM - AN INTRODUCTION
(Micro Scale Experiment)

1. Important Reminders

a. Call students' attention to the fact that even though this is a micro scale experiment, concentrated acids and bases can cause burns in small amounts.

b. Dispense 3 *M* HNO_3, 6 *M* HNO_3, and 6 *M* NH_3, from self dispensing bottles, labeled to show hazards.

c. Demonstration: The reaction of copper with concentrated nitric acid produces NO_2 primarily. A glass ampule can be made by heating the open end of a 25 mm diameter tube with an oxygen-gas torch and drawing a neck down to about 6-8 mm. Cut off the old test tube lip leaving a long neck. Attach the neck to a 3 way stopcock. Attach the other tubes of the stopcock to an aspirator and to a glass tube in a 1-hole stopper. Turn the stopcock to connect the aspirator and the ampule. Evacuate the glass ampule and cool the bottom of the ampule in an icebath. Turn the stopcock to connect the aspirator and the glass tube in a 1-hole stopper. Loosely attach a test tube containing 1 g Cu and 5 mL conc. HNO_3 to the 1-hole stopper. Allow the reaction of Cu/HNO_3 to generate enough NO_2 to flush the air out of the reaction tube. Carefully turn the stopcock so that the NO_2 can enter the evacuated ampule. After the ampule has filled with NO_2, turn the stopcock to again connect the aspirator and the reaction container. Seal the neck of the ampule near the body of the ampule with an oxygen-gas torch.

2. Special Appartus and Equipment

	Per Student
24 well test plate	1
Plastic micropipettes	6
Indicator Color Chart	1 per 2-3 students
Beaker, 500 mL	1
Sealed glass ampule containing N_2O_4/NO_2, (25-50 mL)	1 per class

3. Chemicals

	Per 100 Students
2 *M* NH_3 (aqueous)	700 ml
Universal Indicator solution	200 mL
NH_4Cl, solid	50 g
1*M* HCl	150 mL
2*M* NaOH	150 mL
1*M* $CuSO_4$	150 mL
0.2 *M* $FeCl_3$	200 mL
0.2 *M* $Fe(NO_3)_3$	100 mL
1.0*M* NaCl	100 mL
0.2 *M* KSCN	100 mL
NaF, saturated solution	150 mL
0.2 *M* $K_4[Fe(CN)_6]$	150 mL

0.2 *M* $Pb(NO_3)_2$	150 mL
3 *M* HNO_3	150 mL
0.5 *M* $Cu(NO_3)_2$	150 mL
6 *M* NH_3 (aqueous)	150 mL
6 *M* HNO_3	150 mL

4. Unknowns

None

EXPERIMENT 24
SPECTROPHOTOMETRIC DETERMINATION OF AN EQUILIBRIUM CONSTANT

1. Important Reminders (Students may work in pairs)

   a. If the number of spectrophotometers are limited, a portion of the class may perform this experiment on successive lab periods while the rest of the class is doing other experiments.

   b. Allow about 20 minutes for the spectrophotometers to warm up before calibration.

   c. By using a single cuvette* the student does not have to be concerned about different transmissions of light through different tubes. Be sure the single cuvette is placed in the spectrometer facing the same direction each time.

   d. Dirt and fingerprints will scatter light. Warn the students to keep the cuvette clean.

2. Special Appartus and Equipment

	Per Pair
Burets, 50 mL	2
Burets clamp	1
Spectrophotometer	1 (per 3 pairs)
Cuvette [or matched* small test tubes (5-6 per pair)]	1
Tissues (for wiping cuvettes)	
Beakers (125 mL)	5
Pipet, 5 mL volumetric	1

3. Chemicals

	Per 50 pairs
0.00200 *M* KSCN	3.5 L
0.200 *M* $Fe(NO_3)_3$	3.5 L
0.050 *M* HNO_3	6.0 L
0.00200 *M* $Fe(NO_3)_2$	3.5 L

4. Unknowns

   none

Calibration of ltest tubes used as cuvettes
* Since commercially standardized cuvettes are expensive, an alternative is to prepare standard test tubes by the following method. Select 8-9 test tubes that appear identical. Make a mark on one side at the top. Fill each tube about 3/4 full with a colored solution such as a dilute $CuSO_4$ solution or a dilute ordinary India ink solution. The solution should show an absorbance about the middle of the spectrophotometer scale. Read the absorbance of the solution in each tube so that the mark on each tube faces in the exact same direction in the well. Choose those tubes which have readings which are identical or almost identical to each other. Reject the others.

EXPERIMENT 25
THE SPECTROPHOTOMETRIC DETERMINATION OF THE FORMULA FOR A COMPLEX SPECIES

THE ION (III)-SULFOSALICYLIC ACID COMPLEX

1. Important Reminders

b. Review the article, "Safe Handling of Perchloric Acid in the Laboratory" by Muse, L. A., J. Chem Ed., 1972, 49, A 463. Call students' attention to the hazards involved in the use of $HClO_4$.

c. $HClO_4$ is a **powerful oxidizing agent** and should be handled very carefully. When preparing the solutions for this experiment, clean up all spills quickly. Consider as a **fire or explosive hazard** any organic materials which is exposed to perchloric acid.

d. Be sure that all waste samples are flushed down drains with large quantities of water.

e. *It is recommended that students work in pairs. If they do, the quantities of chemicals can be halved.

2. Special Apparatus and Equipment

	Per Student*
Beakers, 250 or 150 mL	5
Spectrometer	1/10 students
Cuvettes or "matching test tubes"	2 cuvettes or 5-8 matching tubes/10 students
10 mL graduated cylinder or	1
10 mL graduated pipet and bulb	1
Wiping tissues (Kimwipes or similar)	
Graduated cylinder, 100 mL	1

3. Chemicals

Dispense the following reagents from self dispensing bottles (Brinkman, etc.) Label with NFPA hazard code.	Per 100 Students*
Solution A:	
$Fe(NO_3)_3$ in 0.10 *M* $HClO_4$	8.0 L
Solution B:	
0.010 *M* 5-Sulfosalicylic acid in 0.10 *M* $HClO_4$	8.0 L
Solution C:	
4.0 *M* $HClO_4$	2.0 L

4. Unknowns

none

EXPERIMENT 26
SOLUBILITY PRODUCTS OF SLIGHTLY SOLUBLE SALTS

1. Important Reminders

a. Remind students that they must protect their eyes with departmentally approved safety glasses.

b. Remind the students to hold the medicine droppers with the tip pointing down. The solutions should never get in the rubber bulb. They should always leave the eyedropper resting in the beaker with the tip down when they are not using it.

c. K_{sp} for MA should be in the range of 10^{-6} to 10^{-4}. K_{sp} for XY_2 should be in the range of 10^{-7} to 10^{-5} for best results. As the experiment is written, MA is $SrCrO_4$ and XY_2 is $PbBr_2$. However, other choices may be made.

2. Special Appartus and Equipment

	Per Student
50 mL beakers (if not in desks)	4
Medicine droppers (if not in desks)	3
Test tubes	8-9

3. Chemicals

			Per 100 Students
Part A			
	0.20 *M* $Sr(NO_3)_2$	label as $A(NO_3)_2$	600 mL
	0.20 *M* K_2CrO_4	label as K_2X	600 mL
Part B			
	0.20 *M* $Pb(NO_3)_2$	label as $B(NO_3)_2$	600 mL
	0.20 *M* KBr	label as KY	600 mL

4. Unknowns

none

EXPERIMENT 27
NITROGEN AND ITS COMPOUNDS

1. Important Reminders

a. Note and call students' attention to all cautions and to the requirement to wear departmentally approved eye protection at all times in the laboratory.

b. Dispense concentrated HCl, 4 *M* NaOH, and 6 *M* NH_3 from screw-top plastic dropper bottles (Scienceware, etc.) which do not leak when inverted. Dispense 3 *M* HNO_3 from a self dispensing bottle (Brinkman Dispensette, etc.)

c. The gas delivery tubes in Step 1,Step 17, and Step 35 should be preassembled in 1 hole stoppers as a safety precaution.

d. The shape of the gas delivery tube for N_2 is shown in Fig. 27.1 and the tube for ammonia is shown as II.

2. Special Apparatus and Equipment

N_2 Generator (Fig. 27.1)	Per Student
15 cm Pyrex tube (step 1)	1
Beaker, 600-800 mL	1
1-hole stopper with glass delivery tube, preassembled (step 1)	1
1-hole stopper with glass delivery tube, preassembled (step 17)	1
15 cm test tubes	4
Clamp and ring stand	
O_2 Generator (Fig. 27.2)	
2-hole stopper with glass delivery, preassembled	1
tube and eyedropper (Step 35)	1
Test tube, (25 mm x 200 mm)	1
Grease pencil or labels	
Ring stand and utility clamp	1
Beaker 250 mL	1
Solid waste (Cu) receptacle	

3. Chemicals

	Per 100 Students
NH_4Cl, solid	325 g
$NaNO_2$, solid	110 g
Mg, ribbon, 3 cm lengths	100 pieces
CaO, solid	310 g
Wooden splints	400
Litmus paper, red and blue	200 pieces of each
Conc. HCl in dropping bottles	10 mL
0.1 *M* $Cu(NO_3)_2$	1.5 L
4 *M* NaOH in dropping bottles	50 mL
6 *M* NH_3 (aqueous) in dropping bottles	200 mL
6 *M* HCl	200 mL

$(NH_4)_2SO_4$, solid	60 g
Cu, turnings or wire	100 g
3 *M* HNO_3	1 L
Methyl red indicator	50 mL
3% H_2O_2	2 L
3 *M* $FeCl_3$	200 mL

4. Unknowns

none

EXPERIMENT 28
THE ANIONS OF SULFUR
(A Micro Scale Experiment)

1. Important Reminders

a. Note and call students' attention to caution and the need to wear departmentally approved eye protection.

b. Dispense 6*M* H_2SO_4 and 3 *M* HCl from self dispensing bottles (Brinkman Dispensette, etc.).

c. Good students may be given a mixture of two anions*, after they have analyzed a single anion.

2. Special Apparatus and Equipment

	Per Student
24 well test plate	1
Plastic micro pipettes	5-6

3. Chemicals

	Per 100 Students
lead acetate paper	600 strips

Solids (Dispense rice grain sized samples using a soda straw spatula tied to each bottle)

Na_2S	50 g
Na_2SO_3	50 g
$Na_2S_2O_3$	50 g
Na_2SO_4	50 g
Na_2CO_3	50 g

Solutions

3 *M* HCl	400 mL
0.1 *M* $BaCl_2$ (Dispense from dropping bottles).	150 mL
6 *M* H_2SO_4	50 mL
0.05 *M* $KMnO_4$ (Dispense from dropping bottles).	50 mL

4. Unknowns

Issue 0.8 g of one of the solid sodium salts in 3 above to each student as an unknown. 20 g of each salt will be required to prepare 100 unknowns with only 1 anion. For unknowns with 2 anions, mix about 0.5 g of each of two salts and shake in a test tube.

*Recommended pairs for good students :

Na_2S and Na_2SO_4
Na_2SO_3 and Na_2SO_4
Na_2SO_4 and Na_2CO_3
$Na_2S_2O_3$ and Na_2SO_4

EXPERIMENT 29

PREPARATION AND ANALYSIS OF AN INORGANIC COMPOUND

(This experiment may require two laboratory periods to complete.)

1. Important Reminders

a. Call students' attention to the hazards involved in the use of the flammable compounds, ethanol and diethyl ether, and in the use of 15 *M* NH_3, solid KOH, and 6 *M* H_2SO_4. Remind them of the requirement for wearing departmentally approved eye protection at all times while in the laboratory.

b. If this experiment requires two laboratory periods to complete, be sure the inorganic compound is stored in a sealed container with NH_3 vapor.

2. Special Apparatus and Equipment

	Per student
Smooth paper for weighing	10
Buchner funnel and aspirator for vacuum filtration	1
Filter paper	1
Watch glasses	3
Zip-lock plastic bags	1
Erlenmeyer flask, 250 mL	2
Burette, 50 mL	1
Burette clamp	1
Test tubes	3
Beaker, 100 mL	1
Beaker, 250 mL	1

Properly labeled waste containers for diethyl ether and for ethyl alcohol

3. Chemicals

	Per 100 Students
$CuSO_4 \cdot 5H_2O$	600 g
15 *M* NH_3	1 L
95% ethyl alcohol	10 L
Diethyl ether	3 L
KOH, solid pellets	100 g
0.1 *M* $Ba(NO_3)$	20 mL
Starch solution	800 mL
0.08*M* $Na_2S_2O_3$, standardized	6 L
6 *M* H_2SO_4	800 mL
NH_4SCN	250 g
KI	250 g
0.10 HCl, standardized	7 L
Methyl orange indicator (dispense from a dropper bottle)	50 mL
Litmus paper, red	400 pieces

4. Unknowns

None

EXPERIMENT 30
ORGANIC QUALITATIVE ANALYSIS
(Semi-Micro Scale Experiment)

1. Important Reminders
   a. Note and call attention of the students to all cautions including wearing safety glasses even though small amounts of the reagents are used..

   b. You may use petroleum ether or light mineral oil instead methylene chloride, CH_2Cl_2 in step 9.

   c. The waste from the organic reagents should be collected in special waste containers. Consult the Material Safety Data Sheets (MSDS) for appropriate waste disposal methods. DO NOT DUMP THESE CHEMICALS IN THE SINK.

   d. Dispense all organic reagents, bromine-chloromethane, Lucas Reagent, DNPH Reagent, and ceric nitrate reagent from self dispensing bottles (Brinkman Dispensette, Asko SMI, Scienceware) with appropriate labels showing color coded hazaard code labels. (NFPA Hazard Codes). The codes for the particular substance may be obtained from the MSDS sheets.

2. Special Appartus and Equipment

	Per Student
Ceramic 12 well spot plate or micro size test tubes (12)	1
Glass micro (Pasteur) pipets	11
small, micro-sized test tubes	11
20 cm length of nichrome wire	1
cork	1

3. Chemicals

	Per 100Students
cyclohexane	600 mL
cyclohexene	600 mL
phenol	100 mL
ethyl alcohol	600 mL
isopropyl alcohol (2-propanol)	300 mL
tertiary alcohol (2-methyl-2-propanol)	300 mL
propionaldehyde	600 mL
acetophenone	600 mL
hexane	600 mL
*2.0% $KMnO_4$	600 mL
*2 % Br_2 in CH_2Cl_2	450 mL
*5% $FeCl_3$ Reagent	450 mL
*2,4-dinitrophenylhydrazine (DNPH) reagent	600 mL
*Ceric nitrate reagent	600 mL
*Lucas reagent	600 mL

*See preparation section.

4. Unknowns

6 mL/ student

aldehydes
2-ethylhexanal
p-tolualdehyde
ketones
2-butanone
cyclohexanone
propiophenone
acetophenone
phenols
phenol
guaiacol
primary alcohols
hexanol
1-propionol
sec. alcohols
2-or3- pentanol
tert. alcohols
tert-butanol
alkenes
1-octene

Experiment 31
Commercial Easters- Aspirin and Oil of Wintergreen

1. Important Reminders
   a. Note and call students' attention to the hazards involved in the use of concentrated H_2SO_4, 95% ethanol, methanol, and acetic anhydride

   b. Dispense concentrated H_2SO_4, 95% ethanol, methanol, and acetic anhydride from self dispensing bottles (Brinkman Dispensette, etc.). Label with appropriate NFPA hazard codes.

   c. Check MSDS for proper disposal procedures for excess chemicals. The products, aspirin and oil of wintergreen should be collected and not carried out of the laboratory.

   d. The melting point of aspirin may be measured by using the hot oil technique described in the Laboratory Techniques, Section M or by using a commercial device which heats a metal block.

2. Special Apparatus and Equipment

	Per Student
melting point apparatus (hot oil) (heated metal block)	1 per 3 students
Electric hot plate	1 per 8 students
Large ring	
100 mL beaker	1
600 mL beaker	1
125 Erlenmeyer	1
Thermometer	1
Tongs	1
Watch glass	1
Buchner funnel and aspirator	1
Filter paper	2-3 pieces
Ice bath	1 per 3 students

3. Chemicals

	Per 100 students
salicylic acid	350 g
acetic anhydride	700 mL
conc. H_2SO_4	150 mL
95% ethanol	2 L
1% $FeCl_3$	30 mL
methanol	250 mL

4. Unknowns
   none

Experiment 32
Preparation and Properties of Soap, Properties of Detergent

1. Important Reminders
   a. Note and call students' attention to the hazards involved in the use of 50% NaOH, 3 *M* HCl, acetone, cyclohexane, and 95% ethanol. Remind them of the requirement for wearing departmentally approved eye protection at all times in the laboratory.

   b. Dispense 50% NaOH, 3*M* HCl, acetone, cyclohexane, and 95% ethanol from self dispensing bottles (Brinkman, Scienceware, etc.). Label with proper NFPA hazard codes.

   c. For large classes, lard may be purchased from a grocery store. For small classes, "bacon drippings " may be used.

2. Special Apparatus and Equipment

	Per 100 students
universal indicator paper	110 pieces

3. Chemicals

	Per 100 students
Lard	350 g
95% ethanol	900 mL
50% NaOH solution	350 mL
Saturated NaCl solution	3.5 L
3 *M* HCl	600 mL
Oil	60 mL
Acetone	2.5 L
Cyclohexane	150 mL
0.1 *M* $CaCl_2$	700 mL
0.1 *M* $MgCl_2$	700 mL
0.1 *M* $FeCl_2$	700 mL
Detergent solution	3.5 mL

4. Unknowns
   none

Experiment 33
Emission Spectra of Metal Ions
(A Micro Scale Experiment)

1. Important Reminders

   a. Remind students of hazards of working with 6*M* HCl and of looking directly at bright discharge tubes. Departmentally approved eye protection must be worn at all times while in the laboratory.

   b. When all tests have been completed, students should flush all metal ion precipitates into a waste container for safe disposal

2. Special Apparatus and Equipment

Hand held spectrophotometer (Edmund Scientific has some for approximately $100.)	1 per 6 students
Nichrome wire	6-10 pieces
Cork stoppers	6-10 pieces
Test tubes, small	5

Waste containers for metal ions - Check MSDS for proper disposal

3. Chemicals

	Per 100 Students
6*M* HCl Dispense from self dispensing bottles (Brinkman, etc.)	150 mL

Cation Solutions - 1 M -Dispense from plastic dropper bottles. Label with full name of salt.

NaCl	300 mL*
Li_2CO_3	300 mL*
$BaCl_2$	300 mL*
$SrCl_2$	300 mL*

4. Unknowns

   *Same cations as knowns with one cation per unknown. Dispense from plastic dropper bottles. Label with full name of salt.

Experiment 34
Qualitative Analysis: The Analysis of Cation Group I

1. Important Reminders

When all test have been completed, students should flush all metal ion precipitates in a waste container for safe disposal

2. Special Apparatus and Equipment

centrifuge

3. Chemicals

solutions *
1 *M* HCl
1 *M* NH_4I
12*M* HCl
1*M* NH_4NO_3
0.5 *M* K_2CrO_4
15 *M* NH_3
1 *M* $SnCl_2$
6 *M* CH_3COOH
0.2 *M* $K_4[Fe(CN)_6]$
4 *M* KCN
6% H_2O_2
magnesia mixture
5% NaClO
0.2 $HgCl_2$

solids
$Na_2S_2O_4$, solid
iron nail
$H_2C_2O_4 \cdot 2H_2O$

4. Unknowns

Ions of Group II and General unknown

Experiment 35
Qualitative Analysis: The Analysis of Cation Group II

1. Important Reminders

When all test have been completed, students should flush all metal ion precipitates in a waste container for safe disposal. Consult MSDS for safe disposal methods.

2. Special Apparatus and Equipment

centrifuge

Chemicals

solutions	solids
6 *M* HCl	
15 *M* NH_3	$Na_2S_2O_4$, solid
6 *M* NH_3	$H_2C_2O_4 \cdot 2H_2O$
magnesia mixture	
0.2 *M* $HgCl_2$	
6 *M* HNO_3	
5% thioacetamide	
1 *M* $SnCl_2$	
6 *M* CH_3COOH	
0.2 *M* $K_4[Fe(CN)_6]$	
4 *M* KCN	
4*M* NaOH	
6*M* HNO_3	
6 *M* NH_4CH_3COO	

4. Unknowns

Ions of Group II and General unknown

Experiment 36
Qualitative Analysis: The Analysis of Cation Group III

1. Important Reminders

When all test have been completed, students should flush all metal ion precipitates in a waste container for safe disposal. Refer to MSDS from chemical supply companies for safe disposal methods for chemicals.

2. Special Apparatus and Equipment

centrifuge — 1 per 4 students

3. Chemicals

solutions *

dimethylglyoxime
acetone
4 *M* NH_4SCN
4 *M* KF
2 *M* KCH_3COO
6 *M* KNO_2
16 *M* HNO_3
6% H_2O_2
1 *M* $NaNO_2$
aluminum reagent
0.1 *M* $Pb(CH_3COO)_2$
5% $(NH_4)_2S$
6*M* CH_3COOH
6*M* HCl
1*M* HCl
6*M* HNO_3
6*M* NH_3
3*M* NH_3
15*M* NH_3
6% H_2O_2
5% thioacetamide

solids

$NaBiO_3$

4. Unknowns

Metal ions of Group III and General Unknown

Experiment 37
Qualitative Analysis: The Analysis of Cation Group IV

1. Important Reminders

When all test have been completed, students should flush all metal ion precipitates in a waste container for safe disposal. Check MSDS for methods of disposal.

2. Special Apparatus and Equipment

centrifuge	1 per 4 students
Platinum or nichrome wire	1 per 4 students

3. Chemicals

solutions *

1*M* $(NH_4)_2CO_3$
95% ethanol
2*M* $(NH_4)_2C_2O_4$
0.5*M* $(NH_4)_2SO_4$
6 *M* NH_3
3*M* NH_3
12 *M* HCl
6 *M* HCl
16 *M* HNO_3
6 *M* CH_3COOH
6 *M* H_2SO_4
0.1 *M* K_2CrO_4

4. Unknowns

Metal ions of Group IV or General unknown

Experiment 38
Qualitative Analysis: The Analysis of Cation Group V

1. Important Reminders

When all test have been completed, students should flush all metal ion precipitates in a waste container for safe disposal. Consult MSDS for safe disposal techniques.

2. Special Apparatus and Equipment

centrifuge	1 per 4 students
platinum or nichrome wire loops	1 per 4 students

3. Chemicals

6*M* HCl
1 *M* HCl
1.5 *M* Na_2HPO_4
magnesium reagent
saturated $Na_3[Co(NO_2)_6]$
6*M* CH_3COOH
3 *M* CH_3COOH
6 *M* NH_3
16 *M* HNO_3
sodium reagent- saturated $Zn(CH_3COO)_2$ and $UO_2(CH_3COO)_2$

4. Unknowns

Ions of Group V and General unknown

PREPARATION OF SOLUTIONS

All directions are given for ONE LITER of solution. Unless otherwise indicated, the solven[t] is water. In some cases a small amount of the "mother" acid should be added to sal[t] solutions to prevent hydrolysis.

A list of commercial (concentrated) acids and aqueous NH_3 follows. A description of th[e] commercial solution is given together with the amount of commercial solution required t[o] prepare one liter of various concentrations of acids and aqueous NH_3. There is som[e] variation in concentration among batches of commercial solutions, but the variation i[s] usually small.

A list of the other solutions follows the list of commercial acids and aqueous ammonia.

Aqueous NH_3	(15.7 *M*, 29.6% NH_3 Sp. Gr. = 0.901)
6 *M* aq NH_3	382 mL
3 *M* aq NH_3	191 mL
2 *M* aq NH_3	128 mL
1 *M* aq NH_3	64 mL
0.05 *M* aq NH_3	3.2 mL

Acetic acid	(16 *M*, 99.8% CH_3COOH Sp. Gr. = 0.960)
6 *M*	375 mL
3 *M*	188 mL
0.1*M*	6.3 mL
0.05 *M*	3.1 mL
1%	11 mL

Hydrochloric acid	(12.3 *M*, 37.6% HCl Sp. Gr. = 1.190)
6 *M* HCl	488 mL
3 *M* HCl	244 mL
1 *M* HCl	81 mL
0.15 *M* HCl	12.2 mL
0.10 *M* HCl (Standard)	8.1 mL

Nitric acid	(16 *M*, 70.8% HNO_3 Sp. Gr. = 1.423)
6 *M* HNO_3	375 mL
3 *M* HNO_3	188 mL
1 *M* HNO_3	62.5 mL
0.05 *M* HNO_3	3.1 mL

Perchloric acid	
4*M* $HClO_4$	444 mL $HClO_4$ (60%)
1*M*$HClO_4$	111 mL $HClO_4$ (60%)
0.1 *M* $HClO_4$	11 mL $HClO_4$ (60%)

Sulfuric acid	(18.1 *M*, 96.7% H_2SO_4 Sp. Gr. = 1.84)12 *M*
H_2SO_4	663 mL
6 *M* H_2SO_4	331 mL
3 *M* H_2SO_4	166 mL
0.05 *M* H_2SO_4	2.8 mL

Aluminum	Amount per liter
1 *M* $Al(NO_3)_3$	375 g $Al(NO_3)_3 \cdot 9H_2O$
0.1 *M* $Al(NO_3)_3$	37.5 g $Al(NO_3)_3 \cdot 9H_2O$
Aluminum reagent	1 g aluminum (ammonium aurin tricarboxylate)
Ammonium	
1 *M* $(NH_4)_2CO_3$	114 g $(NH_4)_2CO_3 \cdot H_2O$
1 *M* NH_4I	145 g NH_4I
1 *M* NH_4NO_3	80 g NH_4NO_3
2 *M* $(NH_4)_2C_2O_4$	284 g $(NH_4)_2C_2O_4 \cdot H_2O$
2 *M* $(NH_4)_2SO_4$	264 g $(NH_4)_2SO_4$
0.5 *M* $(NH_4)_2SO_4$	66 g $(NH_4)_2SO_4$
1 *M* $(NH_4)_2S$	375 mL $(NH_4)_2S$ sol'n (commercial)
5% $(NH_4)_2S$	250 mL $(NH_4)_2S$ sol'n (commercial)
4 *M* NH_4SCN	304 g NH_4SCN
2 *M* NH_4SCN	152 g NH_4SCN
Barium	
1 *M* $BaCl_2$	244 g $BaCl_2 \cdot 2H_2O$
0.1 *M* $BaCl_2$	24.4 g $BaCl_2 \cdot 2H_2O$
0.1 *M* $Ba(NO_3)_2$	26.1 g $Ba(NO_3)_2$
Basic fuchsin	3 mg/mL in 95% ethanol
Bromine	
2% Br_2/H_2O	20 mL Br_2/L H_2O
2% Br_2/CH_2Cl_2	20 mL Br_2/L CH_2Cl_2 (petroleum ether
Bromothymol blue indicator	0.8 g in 128 mL 0.01 *M* NaOH, dilute to 1 L
Buffer solutions, standard, pH 4, pH 7, and pH 10	purchase ready made

Calcium	Amount per liter
2 *M* $Ca(NO_3)_2$	472 g $Ca(NO_3)_2 \bullet 4H_2O$
1 *M* $Ca(NO_3)_2$	236 g $Ca(NO_3)_2 \bullet 4H_2O$
0.1 *M* $Ca(NO_3)_2$	23.6 g $Ca(NO_3)_2 \bullet 4H_2O$
0.05 *M* $Ca(NO_3)_2$	12 g $Ca(NO_3)_2 \bullet 4H_2O$
Ceric nitrate reagent	With heating, dissolve 40 g ceric ammonium nitrate in a solution of 13 mL conc. nitric acid (caution:burns) in 88 mL. of water.
Cobalt	
0.1 *M* $Co(NO_3)_2$	29 g $Co(NO_3)_2 \bullet 6H_2O$
Copper	
1 *M* $Cu(NO_3)_2$	242 g $Cu(NO_3)_2 \bullet 2\ 1/2\ H_2O$
0.5 *M* $Cu(NO_3)_2$	121 g $Cu(NO_3)_2 \bullet 2\ 1/2\ H_2O$
0.1 *M* $Cu(NO_3)_2$	24.2 g $Cu(NO_3)_2 \bullet 2\ 1/2\ H_2O$
1 *M* $CuSO_4$	250 g $CuSO_4 \bullet 5H_2O$
Crystal violet indicator	0.4 g
Dimethylglyoxime	10 g in 95% ethanol
0.5 *M* Dithiooxamide (alcoholic)	
Diphenylamine sulfonic acid	3.2 g barium diphenylamine sulfonate 5 g Na_2SO_4 . Filter to remove $BaSO_4$
2,4-Dinitrophenylhydrazine (DNPH) reagent	Dissolve 9.6 g DNPH in 48 mL conc. sulfuric acid (caution: burns) . With stirring, add this solution to a solution of 64 mL of water and 224 mL of 95% ethanol (flammable). Filter if cloudy.
Erythrosin B indicator	2 g disodium salt
Fluorescein	3 mg/mL in 95% ethanol
Formic acid, HCO_2H, 0.350 *M*	16.1 g
Formic acid, HCO_2H, 0.300 *M*	13.9 g
6% H_2O_2	(use 6% or dilute 30%: 1 – 5)
3% H_2O_2	(use 3% or dilute 30%: 1 –10)
Iron	
3 *M* $FeCl_3$	711 g $FeCl_3 \bullet 6H_2O$
5% $FeCl_3$ Reagent	33.3 g $FeCl_3 \cdot 6H_2O$ is dissolved in 467 mL water. The reagent should be freshly prepared.

	Amount per liter
1 *M* $FeCl_3$	270.3 g $FeCl_3•6H_2O$
0.1 *M* $FeCl_3$	27.0 g $FeCl_3•6H_2O$
1% $FeCl_3$	1.67 g $FeCl_3•6H_2O$/100 mL solution
0.2 *M* $Fe(NO_3)_3$	81 g $Fe(NO_3)_3•9H_2O$
0.1 *M* $Fe(NO_3)_3$	40 g $Fe(NO_3)_3•9H_2O$
0.0200 *M* $Fe(NO_3)_3$	8.08 g $Fe(NO_3)_3•9H_2O$
0.01 *M* $Fe(NO_3)_3$ in 0.1 *M* $HClO_4$	4.04 g (11.1 mL 60% $HClO_4$)
1 *M* $FeSO_4$	278 g $FeSO_4•7H_2O$
Fehling's solution A	Commercially available
Fehling's solution B	Commercially available
Lead	
0.1 *M* $Pb(CH_3COO)_2$	38 g $Pb(CH_3COO)_2•3H_2O$
1 *M* $Pb(NO_3)_2$	331 g $Pb(NO_3)_2$
0.2 *M* $Pb(NO_3)_2$	66 g $Pb(NO_3)_2$
0.1 *M* $Pb(NO_3)_2$	33 g $Pb(NO_3)_2$
Lucas reagent	With cooling, dissolve 272 g of anhydrous $ZnCl_2$ in 177 mL conc. HCl (caution:burns)
Magnesium	
1 *M* $Mg(NO_3)_2$	256 g $Mg(NO_3)_2•6H_2O$
0.1 *M* $Mg(NO_3)_2$	26 g $Mg(NO_3)_2•6H_2O$
Magnesia mixture	50 g $MgCl_2•6H_2O$ 100 g NH_4Cl 100 mL 15 *M* NH_3
Magnesium reagent	0.1 g p-(nitrobenzeneazo)-resorcinol 0.025 *M* NaOH
Malachite green indicator	1 g
Manganese	
Saturated $MnCl_2$ in 12 M HCl	saturate 12 IM HCl with $MnCl_2$
0.1 *M* $MnSO_4$	15.1 g
Mercury	
0.2 *M* $HgCl_2$	54 g $HgCl_2$
0.05*M* $HgCl_2$	14 g $HgCl_2$
Methyl orange indicator	2 g
Methyl red indicator	0.4 g in 95% ethanol

	Amount per liter
New Methylene Blue indicator	3 mg/mL in 95% ethanol
1*M* oxalic acid, $H_2C_2O_4$	126 g $H_2C_2O_4 \cdot 2H_2O$
0.05 *M* oxalic acid, $H_2C_2O_4$, standardized	6.3 g $H_2C_2O_4 \cdot 2H_2O$
<u>Nickel</u>	
0.1 *M* $Ni(NO_3)_2$	29 g $Ni(NO_3)_2 \cdot 6H_2O$
0.1% Phenol (aqueous)	1 g phenol
Phenolphthalein indicator	10 g in 95% ethanol
<u>Potassium</u>	
2 *M* $KC_2H_3O_2$	196 g $KC_2H_3O_2$
0.2 *M* KBr	24 g KBr
Saturated KCl for glass electrode	available commercially
0.5 *M* K_2CrO_4	97 g K_2CrO_4
0.2 *M* K_2CrO_4	39 g K_2CrO_4
4 *M* KCN	260 g KCN
0.2 *M* $K_4Fe(CN)_6$	84 g $K_4Fe(CN)_6 \cdot 3H_2O$
4 *M* KF	376 g $KF \cdot 2H_2O$
2 *M* KOH	112 g KOH (use <u>lab</u> grade)
0.0400 *M* KIO_3	8.560 g KIO_3
0.4 *M* $KMnO_4$ (sat'd)	65 g $KMnO_4$
0.05 *M* $KMnO_4$	8 g $KMnO_4$
2% $KMnO_4$	20 g $KMnO_4$/980 mL H_2O let stand overnight, and filter into a dark bottle.
0.5% $KMnO_4$	5 g $KMnO_4$/995 mL H_2O
0.250 *M* Potassium acid phthlate, $KHC_8H_4O_4$	51 g
0.3000 *M* Potassium acid phthlate $KHC_8H_4O_4$	61.2 g
1 *M* KNO_3	101 g KNO
0.1 *M* KNO_3	10 g KNO_3
6 *M* KNO_2	511 g KNO_2
0.1 *M* KSCN	10 g KSCN
0.00200 *M* KSCN	0.194 g KSCN

	Amount per liter
<u>Silver</u>	
1 *M* $AgNO_3$	170 g $AgNO_3$
5% $AgNO_3$	50 g $AgNO_3$/950 mL H_2O
0.1 *M* $AgNO_3$	17 g $AgNO_3$
<u>Sodium</u>	
saturated $NaCH_3COO$	saturate 1 L of water with $NaCH_3COO$
0.1 *M* $NaCH_3COO$	8.2 g $NaCH_3COO$
1 *M* NaBr	103 g NaBr
0.1 *M* NaBr	10 g NaBr
sat'd NaCl	370 g NaCl
1 *M* NaCl	58 g NaCl
0.1 *M* NaCl	5.8 g NaCl
sat'd Na_2CO_3	370 g $NaCO_3$
1.5 *M* Na_2CO_3	159 g $NaCO_3$
1.0 *M* Na_2CO_3	106 g $NaCO_3$
sat'd $Na_3[Co(NO_2)_6]$	300 g $NaNO_2$ in 970 mL H_2O 30 mL 16 *M* CH_3COOH 33 g $Co(NO_3)_2 \bullet 6H_2O$ Filter
Detergent Solution(Na dodecyl SO_4) $Na(C_{12}H_{25})SO_4$	50 g "Tide"
Saturated NaF	saturate 1 L with NaF
1 *M* NaF	42 g NaF
0.1 *M* NaF	4.2 g NaF
1.5 *M* Na_2HPO_4	213 g Na_2HPO_4
5% NaClO	Use "Clorox"
50% NaOH	500 g NaOH/500 mL H_2O (<u>Tech</u> grade)6 *M* NaOH
	240 g NaOH
4 *M* NaOH	160 g NaOH
3 *M* NaOH	120 g NaOH
2 *M* NaOH	80 g NaOH
1.5 *M* NaOH	60 g NaOH
0.100 *M* NaOH (standard)	4.00 g NaOH ←
2*M* NaI (or KI)	300g (332 g KI)
1 *M* NaI (or KI)	150 g NaI (166 g KI)
0.1 *M* NaI (or KI)	15 g (17 g KI)
0.1 *M* $NaNO_3$	8.5 g $NaNO_3$
1 *M* $NaNO_2$	69 g $NaNO_2$
1 *M* Na_2SO_4	142 g

34

	Amount per liter
0.1 *M* Na_2S	24 g $Na_2S \cdot 9H_2O$
0.2*M* $Na_2S_2O_3$	50 g $Na_2S_2O_3 \cdot 5H_2O$ (Tech grade)
0.08*M* $Na_2S_2O_3$	19.8 g $Na_2S_2O_3 \cdot 5H_2O$ (Tech grade)
Sodium reagent	120 g $UO_2(CH_3COO)_2 \bullet 2H_2O$ 320 g $Zn(CH_3COO)_2 \bullet H_2O$ 40 mL 16 *M* CH_3COOH, dilute to 1 L. Filter next day.
0.0200 *M* H_2SO_3-starch sol'n	Grind 17 g soluble starch in 50 mL H_2O. Slowly pour paste into 900 mL boiling water. Boil until clear. Add 2.40 g $NaHSO_3$. NO MORE than 12 hours before lab,add mL $6H_2SO_4$ and dilute to 1 L.
starch sol'n	Grind 2 g soluble starch in 30 mL water. Slowly pour slurry into 1 L boiling water. Boil until clear. (Small amounts of starch solution can be prepared by spraying about 5 mL of home "spray starch" in 100 mL H_2O.
<u>Strontium</u>	
1 *M* $SrCl_2$	158.5 g
0.20 *M* $Sr(NO_3)_2$	42 g $Sr(NO_3)_2$
0.010 *M* 5-sulfosalicylic acid	2.54 g SSA•$2H_2O$ in 0.1 *M* $HClO_4$ (11.1 mL 69% $HClO_4$)
<u>Tin</u>	
1 *M* $SnCl_2$	226 g $SnCl_2 \bullet 2H_2O$ in 3 *M* HCl
1 *M* $Sn(NO_3)_2$	19 g
5% Thioacetamide	50 g CH_3CSNH_2/950 mL H_2O
Thymol blue indicator	1 g in 215 mL 0.01 *M* NaOH, dilute to 1 L
Universal Indicator Solution	available commercially
<u>Zinc</u>	
1 *M* $Zn(NO_3)_2$	297 g $Zn(NO_3)_2 \bullet 6H_2O$
0.1 *M* $Zn(NO_3)_2$	30 g $Zn(NO_3)_2 \bullet 6H_2O$
1 *M* Zn SO_4	288 g $ZnSO_4 \cdot 7H_2O$

NITRATE CATION STOCK SOLUTIONS

The indicated quantities are dissolved in enough water (plus HNO_3 in some cases) to give ONE LITER of solution. Note degree of hydration!

STOCK SOLUTIONS are to be diluted 1 – 10 in preparing KNOWNS and UNKNOWNS with the exception of Ba solutions. Barium solutions are to be diluted 1 – 5 ONLY.

Cations	Compound	Amount/L	16 *M* HNO_3
Bi^{3+}	$Bi(NO_3)_3 \cdot 5H_2O$	150 g	100 mL
Cu^{2+}	$Cu(NO_3)_2 \cdot 3H_2O$	225 g	50 mL
Cd^{2+}	$Cd(NO_3)_2 \cdot 4H_2O$	175 g	50 mL
As^{3+}	As_2O_5*	125 g	
Co^{2+}	$Co(NO_3)_2 \cdot 6H_2O$	300 g	50 mL
Ni^{2+}	$Ni(NO_3)_2 \cdot 6H_2O$	300g	50 mL
Fe^{3+}	$Fe(NO_3)_3 \cdot 9H_2O$	375 g	50 mL
Mn^{2+}	$Mn(NO_3)_2 \cdot 6H_2O$	300 g(0.50 pt., 50%)	50 mL
Cr^{3+}	$Cr(NO_3)_3 \cdot 9H_2O$	300 g	50 mL
Al^{3+}	$Al(NO_3)_3 \cdot 9H_2O$	700 g	50 mL
Zn^{2+}	$Zn(NO_3)_2 \cdot 6H_2O$	300 g	50 mL
Ba^{2+}	$Ba(NO_3)_2$	150 g	50 mL
Sr^{2+}	$Sr(NO_3)_2$	150 g	

*Dissolves in hot water to produce H_3AsO_4. Filter to remove traces of insoluble As_4O_6.

CHLORIDE CATION STOCK SOLUTIONS

Cation	Compound	Amount/L	12 M HCL
Bi^{3+}	$BiCl_3 \bullet H_2O$	100 g	100 mL
Cu^{2+}	`$CuCl_2 \bullet 2H_2O$	150 g	50 mL
Cd^{2+}	$CdCl_2 \bullet 2\ 1/2\ H_2O$	125 g	50 mL
Sb^{3+}	$SbCl_3$	125 g	150 mL
Sn^{4+}	$SnCl_4 \bullet 5H_2O$	200 g	150 mL
Co^{2+}	$CoCl_2 \bullet 6H_2O$	240 g	50 mL
Ni^{2+}	$NiCl_2 \bullet 6H_2O$	240 g	50 mL
Fe^{3+}	$FeCl_3 \bullet 6H_2O$	240 g	50 mL
Mn^{2+}	$MnCl_2 \bullet 4H_2O$	200 g	50 mL
Cr^{3+}	$CrCl_3 \bullet 6H_2O$	240 g	50 mL
Al^{3+}	$AlCl_3 \bullet 6H_2O$	600 g	50 mL
Zn^{2+}	$ZnCl_2$	200 g	50 mL
Ba^{2+}	$BaCl_2 \bullet 2H_2O$	150 g	
Sr^{2+}	$SrCl_2 \bullet 6H_2O$	225 g	
Ca^{2+}	$CaCl_2 \bullet 2H_2O$	400 g	
Mg^{2+}	$MgCl_2 \bullet 6H_2O$	550 g	
Na^+	$NaCl$	250 g	
K^+	KCl	300 g	
NH_4^+	NH_4Cl	300 g	

PREPARATION OF KNOWN CATION SOLUTIONS

1. Individual Cations in Groups I, IIA, III

   Dilute 40 mL of each NITRATE CATION STOCK solution to 400 mL.

   The Bi solution should have 40 mL of 16 *M* HNO_3 included in the total volume of 50 mL.

2. Individual Cations in Groups IIB, IV, and V

   Dilute 40 mL of the CHLORIDE CATION STOCK solution to 400 mL Exceptions:
   Dilute 80 mL of Ba CHLORIDE CATION STOCK solution to 400 mL.

   The Sb and Sn solutions should have 40 mL of 12 *M* HCl included in the total volume of 400 mL.

   Use H_3AsO_4 for As solutions. Dilute 40 mL to 400 mL.

3. Group Cation Knowns

a. Group I known	40 mL each of NITRATE STOCK solutions of Bi^{3+}, Cu^{2+}, Cd^{2+} + 40 mL of HNO_3, , and dilute to 400 mL.
b. Group II	40 mL each of CHLORIDE CATION STOCK solutions of Sb^{3+}, Sn^{4+}, and As^{3+} + 40 mL of HCl, and dilute 400 mL. For As^{3+}, use H_3AsO_4 STOCK SOLUTION.
d. Group III known	40 mL each of NITRATE CATION STOCK solutions of Co^{2+}, Ni^{2+}, Fe^{3+}, Mn^{2+}, Cr^{3+}, Al^{3+}, Zn^{2+} + 40 mL HNO_3, and dilute to 400 mL.
e. Group IV known	40 mL each of CHLORIDE CATION STOCK solutions of Sr^{2+} and Ca^{2+}, and 80 mL of Ba CHLORIDE CATION STOCK solution, and dilute to 400 mL.
f. Group V known	40 mL of CHLORIDE CATION STOCK solutions of Mg^{2+}, K^+, Na^+ and NH_4^+, and dilute to 400 mL.

DO NOT MIX CHLORIDE AND NITRATE STOCK SOLUTIONS IF THE RESULTING SOLUTION IS TO BE USED OVER AN EXTENDED TIME.

PREPARATION OF CATION UNKNOWNS

Unknowns are prepared in exactly the same way that Group Cation Knowns are prepared with a few exceptions.

1. Group IIA and B Unknowns

   a. Nitrate unknowns cannot contain Sb^{3+} or Sn^{4+} if they are to be used over an extended time. Hydrous oxides of Sb^{3+} and Sn^{4+} precipitate from nitrate solutions on standing.

   b. Excess "mother" acid may be required to prevent precipitate formation, particularly in solutions that contain Bi^{3+}, Sb^{3+}, or As^{3+}.

2. General Unknowns

   Same exceptions as noted in (1) above.

   A maximum of 10 cations can be included in general unknowns that do not contain Ba. Since the Ba STOCK solution is to be diluted 1 – 5, only 9 cations